The Science of Happiness: Unlocking the Secrets to Joy

Greta Rose

Published by RWG Publishing, 2023.

While every precaution has been taken in the preparation of this book, the publisher assumes no responsibility for errors or omissions, or for damages resulting from the use of the information contained herein.

THE SCIENCE OF HAPPINESS: UNLOCKING THE SECRETS TO JOY

First edition. April 20, 2023.

Copyright © 2023 Greta Rose.

Written by Greta Rose.

Table of Contents

Chapter 1: The Neuroscience of Happiness: How Our Brains Create Joy 1

Chapter 2: The Power of Positive Thinking: Optimism and Happiness 5

Chapter 3: The Science of Gratitude: Cultivating Appreciation for a Happy Life 9

Chapter 4: The Role of Relationships in Our Happiness 13

Chapter 5: The Happiness Equation: Balancing Pleasure and Purpose 17

Chapter 6: The Surprising Benefits of Helping Others: The Connection Between Giving and Happiness 21

Chapter 7: Mindfulness and Happiness: The Art of Being Present 25

Chapter 8: The Joy of Simple Pleasures: Finding Happiness in the Little Things 29

Chapter 9: The Science of Sleep and Happiness: The Connection Between Rest and Joy 33

Chapter 10: The Power of Forgiveness: Letting Go for Greater Happiness 37

Chapter 11: The Psychology of Happiness: Understanding Our Emotional Needs 41

Chapter 12: The Happiness Set Point: Can We Change Our Natural Level of Joy? 45

Chapter 13: The Benefits of Laughter: The Connection Between Humor and Happiness 47

Chapter 14: The Happiness of Human Connection: The Science of Social Support 49

Chapter 15: The Art of Resilience: How to Overcome Challenges and Find Joy 51

Chapter 16: The Happiness of Giving: How Generosity Boosts Our Well-Being 53

Chapter 17: The Happiness of Nature: The Connection Between the Outdoors and Joy 55

Chapter 18: The Power of Positive Relationships: Cultivating Happy Friendships and Romantic Partnerships 57

Chapter 19: The Joy of Meaningful Work: The Connection Between Career and Happiness 59

Chapter 20: The Happiness of Travel: The Science of Exploring the World 61

Chapter 21: The Role of Genetics in Our Happiness: Are Some People Naturally Happier Than Others? 63

Chapter 22: The Science of Self-Compassion: The Art of Being Kind to Yourself 65

Chapter 23: The Benefits of Exercise: How Physical Activity Can Boost Happiness 67

Chapter 24: The Happiness of Creativity: The Connection Between Art and Joy 69

Chapter 25: The Power of Mindset: How Our Beliefs Shape Our Happiness 71

Chapter 26: The Joy of Learning: The Connection Between Knowledge and Happiness 73

Chapter 27: The Happiness of Music: How Sound Can Boost Our Well-Being 75

Chapter 28: The Science of Mind-Body Connection: How Our Physical Health Affects Our Happiness 77

Chapter 29: The Happiness of Authenticity: The Importance of Being True to Ourselves 79

Chapter 30: The Joy of Giving Back: The Connection Between Community Service and Happiness 81

Chapter 1: The Neuroscience of Happiness: How Our Brains Create Joy

The state of mind known as happiness is difficult to pin down and comprehend due to its multifaceted nature. On the other hand, thanks to developments in neuroscience, researchers have made significant headway in deciphering the mysteries behind happiness. The neuroscience of happiness and how our brains create joy will be the topic of discussion in this chapter.

At its most fundamental, experiencing joy is the result of a chemical reaction in the brain. A substance known as dopamine is produced in our brains whenever we have an experience that gives us pleasure, such as eating something delicious or receiving a thoughtful act from a friend. Because it is the neurotransmitter that triggers joy and euphoria, dopamine is often referred to as the "pleasure chemical." This is because it is the chemical that is responsible for the feelings that we experience.

However, the release of dopamine is not the only factor that contributes to happiness. In addition to dopamine and norepinephrine, the brain also produces chemicals like serotonin and oxytocin, both of which are linked to positive emotions. Serotonin is commonly known as the "happy hormone" due to the fact that it helps regulate mood, appetite, and sleep. Oxytocin is commonly referred to as the "love hormone" due to the fact that it is secreted during the process of social bonding and is linked to emotions such as love, trust, and connection.

The structure of the brain, in addition to the chemicals within it, plays an important part in one's level of happiness. Many of the higher-level cognitive functions that are associated with happiness, such as decision-making and goal-setting, are controlled by the prefrontal cortex, which is located at the front of the brain and is responsible for many of the higher-level cognitive functions. The deep-seated region of the brain known as the amygdala is responsible for the processing of feelings and is frequently linked to anxious and fearful states. On the other hand, studies have shown that the amygdala also plays a part in the formation of positive emotions like happiness.

In addition, studies have demonstrated that individuals who report high levels of happiness have higher levels of activity in particular regions of the brain. For instance, the ventral striatum, which is linked to feelings of reward and inspiration, is more active in the brains of individuals who experience frequent bursts of happiness. Similarly, the anterior cingulate cortex, which is associated with empathy and social interaction, is more active in people who are both more socially connected and experience higher levels of happiness. This is because people with these characteristics are more likely to have positive relationships with others.

There are both internal and external factors that can have an effect on a person's level of happiness. While the chemistry and structure of the brain do play a significant role in the creation of happiness, there are also other factors that can have an effect. For instance, studies have shown that the environment we live in, the people we interact with on a regular basis, and the routines we follow can all have an effect on our levels of happiness.

Studies have shown that people who live in urban areas or near green spaces tend to report higher levels of happiness. This could be because of the environment in which they are living. In a similar vein, research has shown that individuals who have access to natural light and outdoor spaces are more likely to experience positive emotions. People who have a greater number of strong social connections have a greater tendency to experience happiness than those who are more socially isolated. One of the most important factors in creating happiness is having strong social connections.

THE SCIENCE OF HAPPINESS: UNLOCKING THE SECRETS TO JOY

Our happiness can also be affected by routines that we perform on a daily basis, such as exercising, getting enough sleep, and practicing mindfulness. Endorphins are natural chemicals that are released into the body during exercise. Endorphins are responsible for producing feelings of happiness and euphoria. A sufficient amount of sleep is not only essential for maintaining good mental health but also has the potential to help regulate mood and lower levels of stress. Meditation and other forms of mindfulness practice such as yoga can help reduce stress and increase feelings of calm and relaxation, both of which can contribute to increased levels of happiness and contentment.

In general, the study of happiness from a neurological perspective is a multifaceted and fascinating subject area that researchers are still actively investigating. Although the chemistry and structure of the brain do play a significant role in the creation of happiness, other factors, such as our environment, the social relationships we have, and the routines we engage in on a daily basis, also play an important part in determining our overall levels of happiness. We can begin to cultivate a greater sense of joy and contentment in our lives by first gaining an understanding of the neuroscience behind happiness.

Chapter 2: The Power of Positive Thinking: Optimism and Happiness

The thoughts we have and the way we react to different circumstances can have a significant effect on the amount of happiness we experience. One way of thinking that has the potential to lead to increased happiness and well-being is optimism, which can be defined as the tendency to focus on positive outcomes and possibilities. In this chapter, we will discuss the power of optimistic thinking and how it can lead to increased levels of happiness.

According to a number of studies, people who have a positive outlook on life tend to report higher levels of happiness. Optimists have a tendency to fixate on positive outcomes and possibilities, which can lead to feelings of hope and motivation in those who are exposed to them. Pessimists, on the other hand, have a propensity to concentrate their thoughts on negative outcomes and possibilities, which can result in feelings of anxiety and depression.

Because it makes it easier for us to deal with challenging situations, optimism is one of the factors that can contribute to increased happiness. Optimists have a tendency to view obstacles and setbacks as temporary and situational rather than as permanent and pervasive. This is because optimists believe that nothing is permanent or pervasive. Even in the face of difficulties, we are able to keep a positive outlook and proceed with our plans with the assistance of this way of thinking.

Being optimistic can also result in increased social support and connections with more people. People are naturally drawn to individuals who are positive and optimistic, and optimists tend to have social networks that are larger and more supportive of them. Having a strong support system can be beneficial in times of stress or crisis, as it can help us cope with difficult situations and maintain a positive outlook on life. This can be beneficial in times of stress or crisis.

Even though optimism is generally regarded as a quality with positive connotations, it is essential to recognize that this quality is not always useful or appropriate. It's possible that having a tendency to ignore or downplay potential risks or challenges, also known as "blind optimism," will result in poor decision-making and outcomes. Similar to how forcing oneself to think positively in the face of overwhelming negativity can be counterproductive and lead to feelings of denial and disengagement, forcing oneself to think positively in the face of overwhelming positivity can be counterproductive.

Thankfully, optimism is a trait that can be taught and improved upon over time. Gratitude is a practice that can help you develop an optimistic outlook on life. We can rewire our brains to focus on the positive aspects of our lives and experiences if we make a habit of bringing our attention to the things for which we are grateful. Mindfulness training, which can assist us in becoming more aware of our own thoughts and feelings and in cultivating a more upbeat perspective, is yet another method for cultivating optimism.

It is equally essential to acknowledge that being upbeat and content are two entirely different states of mind. Even though having a positive outlook and being optimistic can help us deal with difficult situations and keep a positive frame of mind, optimism does not ensure happiness or eradicate unpleasant feelings. Optimism, on the other hand, refers to a way of looking at life that can assist us in locating joy and meaning despite the presence of adversity.

When it comes to increasing one's level of happiness and overall sense of well-being, the power of optimistic thinking and maintaining a positive frame of mind cannot be overstated. Optimists have a tendency to fixate on positive outcomes and possibilities, which can result in feelings of hope, motivation, and resilience in those who are exposed to them. We can develop a more optimistic outlook on life and raise our overall levels of happiness if we practice optimism through gratitude, mindfulness, and positive thinking. This will allow us to experience more joy in our lives.

Chapter 3: The Science of Gratitude: Cultivating Appreciation for a Happy Life

The act of appreciating and expressing thankfulness for the positive aspects of one's life is the essence of the gratitude practice. Research has shown that practicing gratitude not only has numerous benefits for our mental health and happiness but is also an important part of many spiritual and religious practices. Gratitude has been recognized for a long time as an important component of spiritual and religious practices. In this chapter, we will investigate the scientific basis of gratitude and discuss the ways in which developing an attitude of appreciation can lead to a more fulfilling life.

There are a number of positive outcomes that have been linked to gratitude, including increased levels of happiness, improved quality of sleep, and enhanced mental and physical health. Gratefulness may have such a powerful effect on our well-being because it encourages us to concentrate on the positive rather than the negative aspects of our lives. This may be one of the reasons why gratitude is so important. When we train our brains to be grateful for the good things in our lives by practicing gratitude, it can lead to feelings of contentment and satisfaction because it trains our brains to look for and appreciate those good things.

It has been demonstrated through research that expressing gratitude can also have a beneficial effect on the relationships we have. People who are grateful have a greater tendency to be more empathetic and forgiving, which can lead to social connections that are both stronger and more positive. It is possible to strengthen our relationships and experience deeper feelings of love and connection by demonstrating to those around us that we appreciate and value them through the act of expressing gratitude toward them.

Giving thanks has been linked not only to the benefits it provides for our mental and social well-being, but also to the improvements it can bring to our physical health. According to a number of studies, grateful people have a tendency to suffer from fewer physical symptoms and illnesses, have stronger immune systems, and may even have a longer lifespan than their counterparts who are less grateful. One possible explanation for these advantages is that an attitude of gratitude lowers levels of stress, which lowers the risk of adverse effects on our physical health.

The process of cultivating gratitude in our lives can be one that is relatively easy and uncomplicated to undertake. Keeping a gratitude journal in which we record the things for which we feel thankful on a daily basis is one strategy for cultivating an attitude of gratitude. This practice helps us to focus on the positive aspects of our lives and can assist us in shifting our mindset towards appreciation and positivity.

One more technique for developing an attitude of gratitude is to show appreciation to other people. Expressing gratitude toward others can help us build stronger relationships, as well as foster feelings of connection and empathy. This is true whether we are expressing gratitude to a friend for their support or to a coworker for their hard work.

The last thing we can do to cultivate an attitude of gratitude is to simply make the effort to appreciate the less significant aspects of our lives. Taking a moment to appreciate the world around us, whether it be the feeling of the sun on our skin or the splendor of a sunset, can help us cultivate a greater sense of gratitude and joy in our lives. This is true regardless of the experience.

In conclusion, the study of gratitude reveals the numerous advantages that can accrue to those who make an effort to develop an attitude of appreciation in their daily lives. There is a correlation between having a grateful attitude and increased levels of happiness, as well as improved relationships and health. We can shift our mindset toward positivity and create a happier, more fulfilling life by regularly practicing gratitude through activities such as journaling, expressing appreciation to others, and taking the time to appreciate the little things in our lives by taking the time to appreciate the small things in our lives.

Chapter 4: The Role of Relationships in Our Happiness

Because humans are social creatures, the quality of our relationships with other people plays a significant part in determining both our level of happiness and our overall health. In this chapter, we will discuss the significance of relationships in our lives as well as the ways in which the cultivation of positive connections can lead to increased levels of happiness.

Research has shown time and again that one of the most important factors in determining one's level of happiness and well-being is the quality and quantity of one's interpersonal relationships. People who have a greater number of strong social connections tend to have higher levels of happiness, better mental and physical health, and a longer life expectancy than those who are more socially isolated.

One of the many reasons why having healthy relationships is essential to our level of contentment is because they give us a feeling of connection and belonging to a group. When we have healthy relationships with other people, we have a sense that we are valued and supported, which can make it easier for us to deal with difficult situations and stress. On the other hand, when we do not have sufficient social connections, we are more likely to experience feelings of isolation and loneliness, which can be a precursor to feelings of depression and anxiety.

The quality of our interpersonal connections has the potential to influence both our mental and physical well-being. According to a number of studies, individuals who have a strong social support network have a tendency to experience lower rates of heart disease, depression, and cognitive decline. People who have more social connections have a higher likelihood of having better immune function and may even have a longer life expectancy than those who have fewer social connections.

In today's fast-paced and frequently isolating world, cultivating positive connections can be difficult, despite the fact that the benefits of having positive relationships are obvious. On the other hand, there are a few different approaches that we can take in order to construct and preserve positive relationships in our lives.

One tactic is to place a high priority on our relationships and to schedule time in our schedules for social connection. This may involve setting up recurring get-togethers with friends and family, going to various social events, or simply making the effort to catch up with loved ones in-person, over the phone, or through online video chat.

The cultivation of positive qualities in our relationships, such as empathy, kindness, and forgiveness, is yet another tactic that can be utilized. We can strengthen our connections with others and create a social environment that is more positive and supportive by demonstrating empathy toward other people as well as by being kind and forgiving of one another.

Last but not least, it is essential to have the realization that not all relationships are favorable or advantageous. Relationships that are toxic or unhealthy for us can have a significant impact on both our mental and physical health; therefore, it is essential that we establish boundaries and put our own well-being first when we are in such circumstances.

In conclusion, the importance of the roles that relationships play in each of our lives simply cannot be overstated. We are better able to deal with the effects of stress and adversity when we have supportive relationships in our lives. These relationships also help us feel a sense of belonging and connection to others. We can make our social environment more uplifting and satisfying for ourselves and others, as well as raise our overall levels of happiness, if we put an emphasis on the importance of our relationships, work on developing our positive traits, and establish limits when confronted with unfavorable circumstances.

Chapter 5: The Happiness Equation: Balancing Pleasure and Purpose

Many people believe that all it takes to achieve happiness is to simply accumulate more pleasure in one's life. However, studies have shown that there is a lot more nuance to the relationship between pleasure and happiness than what is suggested by this equation. In this chapter, we will discuss the happiness equation and the ways in which a life that strikes a healthy balance between pleasure and purpose can result in a more satisfying existence.

Despite the fact that experiencing pleasure can undeniably boost our levels of happiness, research has shown that doing so on its own is not sufficient to produce happiness that lasts. The process by which we quickly adjust to a positive experience to the point where it no longer brings the same level of pleasure as it did when we were first exposed to it is known as hedonic adaptation. This indicates that the pursuit of pleasure for its own sake can result in a never-ending cycle of wanting more and never feeling satisfied with one's current state.

Instead, the findings of recent studies have pointed to the importance of striking a healthy balance between one's pursuit of pleasure and one's contribution to a greater good. A sense of meaning and direction in our lives is what we mean when we talk about having a purpose, and studies have shown a correlation between having a sense of purpose and higher levels of well-being and overall life satisfaction.

Pursuing activities that provide both pleasure and meaning is one way to strike a balance between the two concepts of pleasure and purpose. For example, volunteering for a cause that is meaningful to us can provide both pleasure and purpose because it allows us to have positive experiences while also bringing a sense of meaning and fulfillment to our lives.

One more way to strike a healthy balance between pleasure and purpose is to zero in on our individual talents and interests. We can instill a sense of purpose and fulfillment in our lives by engaging in pursuits that are congruent with our capabilities and the things that are important to us. For instance, a person who has a strong interest in music might discover that working in the music industry fulfills them while also allowing them to continue to derive pleasure from the activities of making music and listening to music.

It is essential to keep in mind that achieving a healthy equilibrium between gratification and contribution is a uniquely personal and individualized endeavor. It's possible that the things that give one person pleasure and meaning are not the same things that give another person those same feelings. As a result, it is essential to make the effort to consider our individual ideals and interests in order to identify pursuits that satisfy our need for both pleasure and fulfillment.

In addition to recognizing that happiness is not a static state, it is essential to strike a healthy balance between the pursuit of pleasure and the fulfillment of one's responsibilities. It is only natural to go through a variety of feelings and states of mind because life is so dynamic and is full of highs and lows. We can, however, create a life that is more satisfying and meaningful for ourselves by developing a sense of purpose and learning to take pleasure in the simple things that life has to offer.

In conclusion, the formula for happiness involves striking a balance in our lives between doing things that bring us pleasure and doing things that have a purpose. Pursuing pleasure for its own sake can result in hedonic adaptation and a never-ending cycle of wanting more, whereas pursuing purpose can give our lives a sense of meaning and direction. We can cultivate a life that is more fulfilling and meaningful for us if we look for pursuits that offer both pleasure and significance, if we pay attention to our individual capabilities and interests, and if we acknowledge that happiness is not a permanent state.

Chapter 6: The Surprising Benefits of Helping Others: The Connection Between Giving and Happiness

It is a cliché, but it is true that it is more satisfying to give than to receive. Research has shown that helping others can actually have numerous benefits for our own happiness and well-being, despite the fact that this sentiment may sound like a cliche at this point. In this chapter, we will discuss the unexpected advantages that come from helping other people as well as the link that exists between generosity and happiness.

Numerous studies have found a correlation between providing assistance to others and increased levels of happiness and well-being for the helpers. Helping other people gives us a sense of connection and purpose in life, both of which can contribute to a sense of being satisfied and fulfilled in our own lives. This feeling is referred to as the "helper's high," and it is analogous to the rush of endorphins that individuals experience when they exercise.

One of the reasons why helping other people can lead to greater happiness in our own lives is because it can teach us to put the challenges we face in perspective. We can cultivate a deeper sense of gratitude for our own lives and experiences if we direct our attention toward the requirements of others. This has the potential to assist us in experiencing increased contentment and fulfillment with our present situations.

Giving aid to other people can also have a beneficial effect on the relationships we have with other people. We can strengthen our social connections and foster greater feelings of empathy and compassion in ourselves and others by providing support and assistance to other people.

This can result in relationships that are both stronger and more positive, which in turn can contribute to our own happiness and sense of well-being.

Assisting other people can have a beneficial effect not only on our mental and social well-being, but also on our physical health as well. Volunteers and people who engage in other forms of helping behavior tend to have lower rates of heart disease, depression, and chronic pain, according to a number of studies that have been conducted on the topic. Helping other people has also been shown to reduce levels of stress and inflammation, both of which can have a beneficial effect on our overall health.

One of the most unexpected advantages of helping other people is that it can actually give us a greater perception of the abundance of time that is available to us. People who engage in behaviors that help others have a greater tendency to feel as though they have more time available to them, which can lead to greater levels of happiness and well-being for those people. It's possible that this is due to the fact that helping others can give us a sense of purpose and direction, which in turn can make us feel like we have a better handle on our time and priorities.

It is important to note that helping behavior should not be motivated solely by the desire for personal gain, despite the fact that it is obvious that helping others has positive effects on the helper. Helping other people should rather be seen as a sincere expression of compassion and empathy towards those being helped.

To summarize, research has shown that those who give to others tend to report higher levels of happiness. Giving back to the community can improve our own happiness, sense of well-being, as well as our sense of purpose and the way our lives are headed. We can strengthen our social connections, put our own problems in perspective, and even improve our physical health by helping others who are in need of support and assistance. We have the power to transform our lives and the lives of those around us into something more uplifting and satisfying if we work on developing a spirit of compassion and empathy.

Chapter 7: Mindfulness and Happiness: The Art of Being Present

It can be difficult to remain fully present and engaged in our lives in the modern world, which moves at such a breakneck pace and is often frenetic. On the other hand, studies have shown that practicing mindfulness, which refers to the act of being fully present and aware in the here-and-now, can have a variety of positive effects on both our happiness and our well-being. In the following chapter, we will discuss the practice of mindfulness and the ways in which it can lead to increased levels of happiness.

The practice of mindfulness entails paying attention to the here and now while maintaining an attitude that is accepting and free from judgment. When we practice mindfulness, we are able to let go of distractions and worries about the past or the future, which allows us to be fully present in the experiences that we are having. This can result in feelings of calmness and contentment, in addition to a heightened sense of connection to our surroundings and the people who are close to us.

The practice of mindfulness has been shown to have a variety of positive effects on both our mental and physical health, as a result of research. According to a number of studies, practicing mindfulness can ameliorate the signs and symptoms of depressive and anxious disorders, boost cognitive performance, and even bring down the levels of stress hormones in the body. In a similar vein, mindfulness has been connected to improved outcomes in terms of one's physical health, such as lower blood pressure and improved immune function.

Because it enables us to fully experience and appreciate the moment that we are living in, mindfulness can have a significant influence on our level of happiness, which is one of the reasons why it can have such a powerful impact on our lives. When we are mindful, we are able to focus our attention on the here and now, rather than on our worries about the future or our regrets about the past. This can result in a sense of gratitude and appreciation for the fleeting moments of joy and beauty that we encounter throughout our lives.

The practice of mindfulness can also assist us in developing a deeper sense of self-awareness as well as compassion for both ourselves and other people. We can better understand our own needs as well as the needs of those who are close to us if we cultivate a heightened awareness of our thoughts and emotions. This can result in a greater capacity for empathy and compassion toward oneself as well as toward other people, which, in turn, can contribute to our own happiness and sense of well-being.

Meditation is one technique that can be used to develop a more mindful mindset. Sitting still and bringing one's attention to the here and now are the primary goals of meditation. This may be accomplished by concentrating on one's breath, a specific object, or a particular feeling. Both on and off the meditation cushion, cultivating the skill of nonjudgmental presence and presence of mind through regular meditation practice can help us become more adept at these states.

Taking the time to fully engage in the experiences we are having is yet another way to cultivate the practice of mindfulness. Whether we are savoring a delicious meal, taking a walk in nature, or spending time with loved ones, taking the time to fully immerse ourselves in the present moment can help us cultivate a greater sense of mindfulness and appreciation for our lives. Taking the time to fully immerse ourselves in the present moment can help us cultivate a greater sense of appreciation for our lives.

In conclusion, the skill of being present in the here and now and engaging in mindfulness practices can confer a variety of advantages on our happiness and well-being. We can cultivate greater self-awareness, empathy, and appreciation for the fleeting moments of joy and beauty that occur in our lives if we pay attention to the here and now with an attitude that is accepting and does not pass judgment on what we see. We can make our lives happier and more meaningful by committing to a regular mindfulness practice. This will also contribute to an increase in our overall level of happiness.

Chapter 8: The Joy of Simple Pleasures: Finding Happiness in the Little Things

It is not uncommon for members of our consumer-driven and materialistic society to confuse happiness with the acquisition of material goods or the accomplishment of external goals. However, studies have shown that one of the keys to genuine happiness is learning to take pleasure in the more straightforward aspects of one's life. In this chapter, we will discuss the joy that can be found in life's more straightforward pleasures, as well as how discovering happiness in the simplest of things can lead to a life that is more satisfying overall.

The term "simple pleasures" refers to the unremarkable, routine activities that give us a sense of happiness and contentment. These can be as simple as having a cup of coffee in the morning, going for a walk in the great outdoors, or spending quality time with the people you care about. Even though these occurrences might not seem significant on their own, the cumulative effect they have on our happiness and well-being can be significant.

A sense of connection and appreciation for the here and now is one of the benefits that comes from indulging in life's more straightforward pleasures, and it's one of the reasons why doing so can help us feel happier. When we partake in pursuits that bring us pleasure, we are able to give our undivided attention to the activity at hand and become fully immersed in the experience, which often results in a sense of contentment and satisfaction.

Research has also demonstrated that indulging in life's more straightforward pleasures can assist us in developing a more upbeat and appreciative frame of mind. We can retrain our brains to concentrate on the positive rather than the negative aspects of the experiences we have if we take the time to appreciate the simple pleasures that are a part of our everyday lives. It's possible that doing this will make you happier and give you a more positive outlook on life.

Finding happiness in the little things in life can have a beneficial effect not only on our mental well-being but also on our physical health. This is because taking pleasure in the experiences that life has to offer can help reduce stress and anxiety. According to a number of studies, people who participate in activities that they find enjoyable on a regular basis tend to have lower levels of stress and inflammation in their bodies, both of which can contribute to improved overall physical health.

Participating in pursuits that are congruent with one's core values and areas of interest is one strategy for rediscovering joy in life's more unpretentious pleasures. We have the power to imbue our lives with more significance and contentment simply by engaging in pursuits that bring us pleasure and gratification. For instance, a person whose passion is gardening may find joy in tending to their plants and watching them grow, whereas a person whose passion is music may find joy in attending concerts or playing an instrument. Both of these activities allow the individual to pursue their passion.

One more way to discover happiness in life's uncomplicated pleasures is to cultivate an attitude of gratitude and appreciation for the fleeting moments of joy and beauty that occur in each of our lives. This can be as easy as pausing for a brief period of time to admire a breathtaking sunset or to savor the flavor of a mouthwatering meal.

In conclusion, the contentment that comes from life's more uncomplicated pleasures can have a significant bearing on our overall happiness and sense of well-being. We can cultivate a more optimistic and grateful mentality, strengthen our connections to the here and now, and even improve our physical health outcomes by finding joy in the small, everyday experiences that life has to offer. We can make our lives happier and more meaningful by partaking in pursuits that are congruent with our core beliefs and pursuits, as well as by pausing to savor the simple pleasures and breathtaking sights that occur naturally throughout our days.

Chapter 9: The Science of Sleep and Happiness: The Connection Between Rest and Joy

Many people fail to recognize the significance of sleep as a contributor to their overall happiness and well-being. On the other hand, studies have shown that getting an adequate amount of restful sleep is not only important for our mental and physical health but also has the potential to have a significant impact on our levels of happiness. In this chapter, we will investigate the scientific aspects of happiness and rest, as well as the connection that exists between the two.

Numerous pieces of research have converged on the conclusion that getting enough high-quality sleep is essential to our overall mental and physical health. When we are asleep, a number of important processes take place in both our bodies and our brains. These processes include the consolidation of memories, the repair of damaged tissue, and the regulation of hormones. This indicates that getting sufficient amounts of restful sleep is necessary if one wishes to maintain optimal levels of both physical and mental function.

In addition to the positive effects that it has on our physical and mental health, sleep is also critical to the maintenance of our mental and emotional wellbeing. According to a number of studies, people who get an adequate amount of sleep have a tendency to have better control over their emotions and a more upbeat disposition, whereas those who don't get enough sleep are more likely to experience negative emotions and erratic moods.

One of the many reasons why getting enough sleep is critical to our mental health is because it assists in maintaining our brain's chemical balance. Neurotransmitters like serotonin and dopamine, which are associated with a happy mood and a sense of well-being, are released into the bloodstream by our brains while we sleep. In contrast, if we don't get enough sleep, our brains produce stress hormones like cortisol, which can make us feel irritable and cause mood swings. Sleep deprivation also makes us more prone to experiencing negative emotions.

One more reason why getting enough sleep is essential to our mental health is that it gives us a feeling of revitalization and fresh start each morning when we wake up. When we awake in the morning feeling rested and refreshed, we are in a better position to face the challenges of the day with a positive and optimistic frame of mind.

In spite of the abundant evidence demonstrating the positive effects of sleep, it can be difficult to get a sufficient amount of restful sleep in today's fast-paced and stressful world. However, there are a few different approaches that we can take to enhance both the quantity and quality of the sleep that we get.

Developing a regular pattern for when you go to bed and wake up is one tactic. Our circadian rhythms can be better regulated, and the quality of our sleep can be improved, if we maintain a consistent bedtime and wake time throughout the week. In a similar vein, developing a soothing ritual to follow before going to bed, such as soaking in a warm bath or reading a book, can assist us in winding down and getting ready for a restful night's sleep.

Creating an atmosphere that is conducive to restful sleep is yet another tactic. This can include making sure that our bedroom is cool, dark, and quiet, as well as avoiding stimulating activities such as watching television or using electronic devices in the hours leading up to bedtime.

In conclusion, it is essential to acknowledge that the process of sleeping is deeply personal and unique to each person. It is essential to try out a variety of approaches and pay attention to our own bodies in order to determine which ones are most effective for us. This will allow us to figure out what works best for us individually.

In conclusion, it is abundantly clear from the research that has been conducted on the topic of sleep and happiness that getting an adequate amount of high-quality sleep is not only necessary for our mental and physical well-being, but also has the potential to have a significant impact on our levels of happiness. We can improve the quality and quantity of our sleep, as well as create a life that is more fulfilling and joyful, if we establish a consistent sleep routine for ourselves, create an environment that is conducive to sleep, and pay attention to our own bodies.

Chapter 10: The Power of Forgiveness: Letting Go for Greater Happiness

The capacity to forgive others is a potent tool for enhancing both our happiness and our well-being. When we harbor resentment, anger, or bitterness toward other people, it has the potential to make us feel burdened and prevent us from experiencing joy and fulfillment in our lives. In the following chapter, we will discuss the efficacy of forgiving others and how learning to let go can ultimately lead to more happiness.

Letting go of negative feelings we have toward someone who has wronged us is an essential part of the forgiveness process. This can be a challenging process, particularly if the damage caused was severe or will continue for some time. However, studies have shown that being able to forgive someone can have a number of positive effects on our mental and physical well-being.

Forgiveness enables us to let go of negative feelings and thoughts that have the potential to hold us back, which is one of the reasons why it is such a powerful concept. When we harbor anger or resentment toward another person, we frequently find that we are mentally reliving an experience that has caused us pain over and over again. This can leave us feeling drained and exhausted, and it can also prevent us from participating fully in our lives and the experiences we have.

The act of forgiving a wrongdoing can also lead to improved mental health. According to a number of studies, individuals who are able to forgive others tend to experience lower levels of stress and anxiety, and they are also less likely to exhibit symptoms of depression. In a similar vein, forgiving someone has been found to be associated with increased feelings of empathy, compassion, and gratitude, all of which can contribute to a life that is more upbeat and fulfilling.

Forgiveness can have beneficial effects not only on our mental health but also on our physical health, in addition to the benefits it already provides for our mental health. According to studies, people who bottle up their anger and resentment for a long time tend to have higher levels of inflammation in their bodies, which can contribute to a variety of physical health issues. On the other hand, it has been found that forgiveness is associated with reduced levels of inflammation and improved overall physical health.

Reappraisal of one's cognitive processes is one technique that can be utilized in the process of forgiving another. This requires recasting the circumstance in a more favorable light and centering our attention on the positive qualities possessed by the individual who has wronged us. For instance, rather than focusing on the hurtful actions of a friend, we could instead concentrate on the positive qualities that they possess and the enjoyable experiences that we have had in the company of that friend.

One more method of forgiving others is to exercise empathy and compassion toward those who have wronged you. We can cultivate greater empathy and compassion toward the person who has wronged us if we make an effort to understand the experiences and motivations of the person who has caused us harm. This can help us see the situation in a way that is more nuanced and understanding, and it can lead to greater forgiveness and emotional healing for us as well.

In conclusion, it is unmistakable that forgiveness possesses tremendous power. If we are able to let go of the negative feelings and thoughts we have toward other people, we will be able to create a life that is more positive and fulfilling. We can foster a greater capacity for forgiveness and emotional well-being in ourselves by engaging in cognitive reappraisal, practicing empathy and compassion, and listening to the perspectives of others. We have the ability to make our lives as well as the lives of those around us more joyful and fulfilling if we accept forgiveness as a tool for personal development and physical healing.

Chapter 11: The Psychology of Happiness: Understanding Our Emotional Needs

The feeling known as happiness is intricate and multidimensional, and it is impacted by a wide variety of internal and external factors. In this chapter, we will discuss the psychological aspects of happiness as well as the emotional requirements that are associated with maintaining a healthy lifestyle.

At its most fundamental level, happiness can be defined as a state of having positive emotions and being in good health. However, studies have shown that happiness is not a fixed state but rather a fluid and ever-evolving process rather than a state in and of itself. Our levels of happiness are affected by a wide variety of internal and external factors, such as our genetics, personality characteristics, social support networks, and the circumstances of our lives.

The fulfillment of our emotional needs is an essential component of happiness. When we talk about our emotional needs, we're referring to the psychological and social requirements for achieving our full potential in terms of well-being. These requirements can include things like love, acceptance, safety, and the ability to make one's own decisions.

According to the findings of numerous studies, it is essential for our overall happiness and well-being to have our emotional requirements fulfilled. We have a tendency to feel more positive emotions, have greater life satisfaction, and experience fewer symptoms of mental health problems like depression and anxiety when our emotional needs are met.

The need for social support is one of the more intangible aspects that contributes significantly to our level of happiness. Studies have shown time and again that individuals who have robust social support networks have a tendency to report higher levels of happiness and experience greater well-being than those individuals who are socially isolated. There are many different groups and individuals that can provide social support, such as family, friends, romantic partners, and community organizations.

Having a sense of purpose and meaning in life is an additional emotional requirement that is essential to our happiness. When we believe that our lives have a reason for being, we have a tendency to experience a greater sense of fulfillment and contentment with our existence. This can come from a variety of places, such as one's job, their hobbies, or the relationships they have.

Last but not least, independence is an essential psychological requirement for our happiness. The term "autonomy" refers to our capacity to make decisions and carry out activities in a manner that is congruent with our own ideals and objectives. We have a tendency to feel a greater sense of happiness and well-being when we have the perception that we have control over our lives and when we are able to pursue our own individual interests and passions.

It is essential for us to engage in activities and patterns of behavior that are congruent with our ideals and objectives if we hope to satiate our emotional requirements and enhance our levels of happiness. This could involve spending time with loved ones, engaging in meaningful work, pursuing hobbies and interests, looking for new experiences and challenges, or any combination of the above.

In conclusion, the psychology of happiness is a complex and multifaceted field, which is influenced by a variety of factors both internal and external to the individual. It is possible for us to live lives that are more satisfying and joyful if we take the time to become aware of our emotional needs and to engage in pursuits and activities that are congruent with our values and objectives. We have the ability to increase the amount of happiness in our lives as well as create a more positive and satisfying existence if we place an emphasis on the importance of social support, meaning and purpose, and autonomy.

Chapter 12: The Happiness Set Point: Can We Change Our Natural Level of Joy?

The term "happiness set point" refers to the idea that every person has a "natural level of happiness" that is largely determined by their genetics as well as the conditions that they face in their lives. This may appear to be a pessimistic outlook on happiness; however, research has shown that we may be able to increase our levels of happiness through intentional effort and changes in behavior. In this chapter, we will discuss the idea of the happiness set point, as well as the question of whether or not we are able to alter our baseline levels of happiness.

The happiness set point is a conceptual framework that proposes that every person possesses an innate level of contentment that they have a propensity to return to over the course of their lifetime regardless of the conditions in which they find themselves. It is believed that genetics, personality traits, and experiences in one's formative years play a significant role in determining this set point.

The concept of a happiness set point may be discouraging to some people, but research has shown that we may be able to raise our levels of happiness by making conscious efforts and making adjustments to our behaviors. For instance, research has shown that engaging in activities that bring us joy and fulfillment, such as spending time with loved ones, pursuing hobbies and interests, and volunteering, can lead to greater levels of happiness and life satisfaction in the long run.

In a similar vein, studies have shown that cultivating an attitude of gratitude and mindfulness can assist us in developing a perspective on life that is more upbeat and optimistic. We can retrain our brains to concentrate on the positive rather than the negative aspects of the experiences we have by concentrating on the here and now and by appreciating the small moments of joy and beauty that occur throughout our lives.

Increasing the amount of kindness and generosity that you show to other people is yet another method for achieving greater levels of happiness. Studies have shown time and again that people who engage in acts of kindness and generosity have a tendency to be happier and experience greater life satisfaction than those who do not engage in such behaviors. It's possible that this is because helping other people can provide us with a sense of purpose and a connection to something that's much bigger than ourselves.

Research has shown that making conscious efforts and making adjustments to our behaviors can have a significant impact on the amount of happiness we experience, even though it may be difficult to alter our baseline level of contentment. We can increase our levels of happiness and create a more positive and fulfilling existence by engaging in activities that align with our values and goals, practicing gratitude and mindfulness, and practicing kindness and generosity towards others.

In conclusion, the idea that an individual possesses a natural level of happiness that is largely determined by their genetics and the circumstances of their life is what is meant by the concept of a happiness set point. Research has shown that we may be able to increase our levels of happiness by making conscious efforts and modifying the behaviors that we engage in. While this may appear to be discouraging, the research has shown that this may be possible. We can increase our levels of happiness and create a more positive and fulfilling existence by engaging in activities that bring us joy and fulfillment, practicing gratitude and mindfulness, and practicing kindness and generosity towards others.

Chapter 13: The Benefits of Laughter: The Connection Between Humor and Happiness

The expression "the best medicine" is frequently applied to laughter, and there is a good reason for this. Research has shown that humor and laughter have a number of positive effects on our mental and physical health, and they also have the potential to have a sizeable influence on the degree to which we experience joy. The purpose of this chapter is to investigate the positive effects of laughter as well as the relationship that exists between humor and contentment.

One of the most important advantages of laughing is that it has the power to relieve stress and tension. When we laugh, the endorphins in our bodies release, which are natural chemicals that promote feelings of pleasure and overall well-being. When we laugh, endorphins are released. This has the potential to make us feel more relaxed and at ease, and it also has the potential to counteract the negative effects that stress has on our bodies and minds.

Laughter has been shown to have a variety of positive effects on our physical health, in addition to the stress-relieving benefits that it provides. Laughter has been shown to improve immune function, lower blood pressure, reduce pain and inflammation in the body, and all of these benefits have been backed up by scientific research. Laughter, in a similar vein, has been associated with better cardiovascular health, including a lower risk of heart disease and stroke.

Laughter and humor can also have a significant positive effect on the mental health of individuals. When we laugh, we feel a surge of joy and positivity that can help us fight off negative feelings such as anxiety and depression. Laughter, in a similar fashion, has been shown to foster feelings of connection and social support, both of which are essential to our mental health.

The application of humor therapy is one method that can be utilized to bring more humor and laughter into our everyday lives. The use of humor and laughter as a therapeutic tool to promote mental and physical well-being is at the heart of this approach. Laughter yoga, comedy improv, and humorous storytelling are just a few examples of the many applications that can be found in the field of humor therapy.

One more thing we can do to bring more humor into our lives is to be on the lookout for instances in which we can laugh and be playful during the course of our daily activities. This might involve watching a funny movie or show on television, spending time with friends who are able to make us laugh, or engaging in activities that are goofy and playful with the people we care about.

To summarize, there is no doubt that laughter has a number of positive effects. Humor and laughter provide a multitude of benefits for our well-being and happiness, ranging from the alleviation of stress and tension to the enhancement of both physical and mental health. We can cultivate a more positive and joyful existence by increasing the amount of laughter and playfulness in our lives. This can be accomplished through the use of humor therapy or simply by engaging in playful activities with the people we care about.

Chapter 14: The Happiness of Human Connection: The Science of Social Support

Connecting with other people is an essential part of our lives and plays an important part in determining our level of contentment and overall health. Research has consistently shown that having strong social support networks can have numerous benefits for our mental and physical health, and can contribute to greater happiness and a greater sense of satisfaction with one's life. In this chapter, we will discuss the importance of human connection to our happiness as well as the science behind the concept of social support.

The emotional, informational, and practical assistance that we get from other people is what's known as "social support," and it's provided by other people. This can come from a wide variety of places, such as the family, friends, romantic partners, and community groups you're involved in. The provision of emotional support during times of stress or crisis, the offering of advice or direction, and the provision of practical assistance such as assistance with chores or transportation are all examples of the many manifestations that social support can take.

According to the findings of numerous studies, having a strong social support system can have a positive impact on both our mental and physical health. People who have strong social support networks typically have lower rates of stress, anxiety, and depression, as well as a lower risk of developing physical health problems such as cardiovascular disease and chronic pain.

Because it fosters a feeling of connection and belonging, having strong social support is critical to our overall health and happiness for a number of reasons. It is much more likely that we will feel positive emotions such as love, empathy, and compassion when we have a strong sense of connection to other people. In a similar vein, having social support can provide a sense of purpose and meaning, because it gives us the impression that we are a part of something that is bigger than ourselves.

One more reason why having social support is essential to our health is that it acts as a defense mechanism against the negative effects of stress. Having the support of others can help us cope with challenging or stressful situations more effectively and reduce the negative impact that stress has on our bodies and minds. When we are faced with challenging or stressful situations, having the support of others can help us.

It is essential to place a high value on the relationships with other people in our lives if we wish to develop robust social support networks. This could include spending time with loved ones, taking part in community groups or clubs, or volunteering for a cause that is important to us. In a similar vein, it is essential to develop empathy and compassion toward other people because doing so can assist us in developing deeper and more meaningful connections with the people in our immediate environment.

In conclusion, it is abundantly clear from the research that has been conducted on the topic of social support that having strong social support networks is essential to our mental and physical well-being, and that it can contribute to greater levels of happiness and life satisfaction. We can make our lives happier and more fulfilling by making human connection a higher priority in our daily activities. This could take the form of prioritizing time spent with loved ones, involvement in community organizations, or the cultivation of empathy and compassion for the plights of others.

Chapter 15: The Art of Resilience: How to Overcome Challenges and Find Joy

Our capacity to face the trials and tribulations of life with fortitude and perseverance can have a significant bearing on the degree to which we experience happiness and flourish in life. In this chapter, we will discuss the art of resilience as well as the ways in which we can cultivate resilience in order to conquer obstacles and discover joy in our lives.

The capacity to respond effectively to challenging circumstances and emerge stronger as a result is known as resilience. When we are confronted with obstacles or failures, resilience enables us to pick ourselves up, dust ourselves off, and carry on regardless of the difficulty of the situation. As a result of the fact that it enables us to navigate the highs and lows of life with greater ease and determination, research has shown that resilience is an important factor in determining one's level of happiness and well-being.

The development of a growth mindset is an essential component of resilience training. A growth mindset is a way of thinking that emphasizes the potential for growth and learning, even in the face of challenges and setbacks. This is in contrast to a fixed mindset, which places less emphasis on the potential for growth and learning. When we adopt what is known as a growth mindset, rather than viewing challenges as insurmountable obstacles, we are able to see them instead as opportunities for growth and learning.

Creating a sense of meaning and purpose in one's life is another essential component of developing resilience. We are better able to navigate challenges and setbacks with resilience and determination when we have a clear sense of the values and goals we aspire to achieve in our lives. This may entail engaging in activities that are congruent with our values and sense of purpose, or it may involve setting goals that are attainable, realistic, and attainable.

In a similar vein, developing an attitude of gratitude and appreciation for the positive aspects of our lives can assist us in enhancing our resilience and locating joy even in the face of challenging circumstances. We can cultivate a greater sense of resilience and optimism within ourselves if we train our attention on the positive aspects of the experiences we've had.

It is essential to place a high priority on practicing self-care and self-compassion in order to develop resilience and be successful in overcoming obstacles. This could mean engaging in activities that are beneficial to both one's physical and mental well-being, such as going to the gym, practicing meditation, or participating in therapy. In a similar vein, it is essential to treat oneself with kindness and compassion, as well as to engage in activities that promote self-forgiveness and self-acceptance.

To summarize, the ability to bounce back quickly from adversity is an indispensable quality for maintaining our happiness and well-being. We are able to navigate the challenges of life with greater ease and determination if we cultivate a growth mindset, develop a sense of purpose and meaning, practice gratitude and appreciation, and prioritize self-care and self-compassion. In spite of the challenges we face, we have the ability to experience happiness and fulfillment if we master the art of resilience.

Chapter 16: The Happiness of Giving: How Generosity Boosts Our Well-Being

Research has shown that being generous can have numerous positive effects on our mental and physical well-being. Generosity is one of the most fundamental aspects of human nature. In this chapter, we will discuss how generosity can improve our well-being as well as the happiness that comes from giving to others.

There are many different ways that we can give to others, such as donating to charitable organizations or volunteering our time and resources. Research has shown time and again that people who engage in acts of generosity tend to experience greater levels of happiness and life satisfaction than those who do not engage in such behaviors.

Giving to others fosters feelings of connection and social support, which is one of the many reasons why it is so important for our overall well-being. Giving to others allows us to connect with them on a deeper level, and this can help foster a sense of community and belonging when combined with the act of giving itself. Giving to others can also give us a sense of purpose and meaning because it allows us to feel as though we are contributing to something that is more significant than ourselves.

Giving to others is beneficial to our mental health for a number of reasons, one of which is that it can increase feelings of gratitude and appreciation. We are able to see the positive impact that our actions have on the world around us when we give to others, and this can help us cultivate a greater sense of gratitude and appreciation for the good things that are a part of our lives.

Giving to others has been shown to improve our physical as well as mental health, according to a number of scientific studies. Generosity is associated with lower levels of stress and anxiety, as well as possibly improved immune function and cardiovascular health for the giver. People who engage in generous acts tend to give more.

It is essential to place a high priority on generosity as a fundamental principle if we wish to foster a greater sense of generosity in our daily lives. This could mean giving our time and resources to causes that are important to us, making financial contributions to charitable organizations, or simply showing kindness and compassion to the people in our immediate environment.

Giving back to the community not only improves our own well-being, but it also has the potential to make the world we live in a better place. We can cultivate a society that is more upbeat and compassionate if we make a meaningful contribution to the health and happiness of those around us, and we can also make our own lives more fulfilling by doing so.

In conclusion, there is no question that giving brings happiness. We can both improve our own well-being and contribute to the creation of a world that is more upbeat and compassionate if we engage in acts of generosity. We can cultivate greater happiness, connection, and purpose in our lives if we make generosity a central value in our lives and make it a priority.

Chapter 17: The Happiness of Nature: The Connection Between the Outdoors and Joy

Spending time outdoors has been shown to have numerous benefits for our happiness and overall life satisfaction, and research has shown that spending time in nature has a significant impact on both our mental and physical well-being. In this chapter, we will discuss the joy that can be found in nature and the connection that exists between being outside and experiencing happiness.

The term "spending time in nature" can refer to a wide variety of activities, such as hiking, camping, gardening, or even just sitting outside and taking in the natural world that surrounds us. Researchers have found time and again that people who spend time outdoors report higher levels of happiness and overall life satisfaction compared to those who do not spend time outdoors.

One of the many reasons why spending time in nature is so essential to our overall health is because it helps us feel more at ease and relaxed. Spending time in natural environments allows us to detach ourselves from the pressures and responsibilities of daily life, which in turn can foster feelings of peace and serenity in us when we return home.

Spending time outside in natural settings can also foster feelings of connection and awe in people. We are able to make a connection with something that is much bigger than ourselves when we are surrounded by the splendor and majesty of the natural world, and this can contribute to the development of a sense of awe and appreciation.

The ability of nature to promote both mental and physical health and well-being is yet another reason why it is essential for our well-being. Time spent outside has been associated with lower levels of stress, anxiety, and depression, and it may also have a positive impact on immune function and cardiovascular health. This is particularly true for people who spend more time outdoors.

Prioritizing time spent outside and actively seeking out opportunities to connect with the natural world are two of the most important things we can do to bring more of the natural world into our everyday lives. This could involve going for a walk every day in a nearby park, organizing recurring trips to go camping or hiking, or even just spending time in our own backyard or on our balcony.

When you spend time outside, it is equally important to cultivate a sense of mindfulness and presence in order to get the most out of the experience. We can foster a stronger sense of connection and appreciation for the natural world around us, as well as feelings of calm and relaxation, by bringing our attention to the sights, sounds, and sensations that are present in that natural world.

In conclusion, it is undeniable that nature brings happiness. Spending time in natural settings can not only improve our mental and physical health but also help us develop a stronger sense of connection and awe in the world around us. We can create a life that is more satisfying and brings us more joy if we make spending time outside a priority and work on developing a sense of mindfulness and presence.

Chapter 18: The Power of Positive Relationships: Cultivating Happy Friendships and Romantic Partnerships

Research has shown that cultivating healthy and happy relationships can have numerous benefits for both our mental and physical health. A positive relationship with others is an essential component of our happiness and well-being. In this chapter, we will discuss the power of healthy relationships as well as the ways in which we can cultivate friendships as well as romantic partnerships that are filled with joy.

Feelings of trust and respect for one another, as well as support and encouragement, are hallmarks of healthy relationships. It is possible for us to experience greater happiness and life satisfaction when we have positive relationships with other people, and we may also have better mental and physical health outcomes as a result of these relationships.

Prioritizing communication that is both open and honest is one strategy for cultivating positive relationships, which is another strategy. When we are able to communicate our thoughts, feelings, and needs to those around us in a way that is both open and respectful, we are able to form connections that are both deeper and more meaningful with those we are surrounded by.

In a similar vein, it is essential to work on developing feelings of empathy and compassion toward other people. We can strengthen the bonds of trust and respect in the relationships we have with those we care about by making an effort to comprehend and acknowledge the viewpoints and experiences of those we hold dear.

In addition to developing healthy friendships and romantic partnerships, it is essential to establish boundaries and put our own wants and needs first in order to ensure that we are taking care of our own health. This could mean refusing to participate in requests or activities that are not congruent with our core beliefs or requirements, or it could mean scheduling in time for self-care and relaxation.

The capacity to forgive one another and make amends following a disagreement or misunderstanding is an additional quality that contributes significantly to the quality of positive relationships. When we are able to work through conflict and repair relationships in a way that is healthy and constructive, we are able to cultivate connections with those around us that are both deeper and more resilient.

In conclusion, there is no denying the power that comes from having positive relationships. We can improve our mental and physical well-being, as well as create a life that is more fulfilling and joyful, by cultivating friendships and romantic partnerships that are healthy and happy for all parties involved. We are able to build stronger and more resilient connections with the people around us if we put an emphasis on open and honest communication, empathy and compassion, establishing healthy boundaries and taking care of ourselves, and forgiving and making amends.

Chapter 19: The Joy of Meaningful Work: The Connection Between Career and Happiness

Finding meaningful work can have numerous positive effects on an individual's happiness and overall well-being, as indicated by a number of studies. Our professions and work lives are an essential component of our identities and purposes. In this chapter, we will discuss the happiness that can be attained through meaningful work, as well as the connection between one's vocation and overall happiness.

Work that is meaningful to us is work that is in line with our core values and sense of purpose, as well as work that allows us to feel fulfilled and satisfied in our work. It is possible for us to experience greater happiness and life satisfaction when we are engaged in meaningful work, and we may also have better mental and physical health outcomes as a result of this engagement.

One of the many reasons why engaging in meaningful work is essential to our overall health and happiness is because it gives us a sense of purpose and direction in life. It is possible for us to experience a greater sense of fulfillment and contentment in our lives when the work that we do is in line with our beliefs, principles, and sense of purpose.

Work that is meaningful to one's life also has the potential to foster a sense of connection and community. We are better able to connect with others who share our interests and passions when we are engaged in work that is in alignment with our values and purpose. This can help to foster a sense of belonging and connection among those who are involved.

Work that is meaningful to us can also foster feelings of mastery and accomplishment, which is another reason why it is important for our well-being for us to do work that is meaningful to us. We are able to feel a sense of accomplishment and satisfaction when we are working at a job that pushes us beyond our comfort zone and gives us the opportunity to learn new skills and improve our existing ones.

In order to find work that is meaningful to us, it is important to take some time to consider our core beliefs, interests, and passions, and then look for opportunities that are congruent with these facets of our personhood. This could mean pursuing additional education or training, looking into new career paths or industries, or working to develop new skills or competencies.

In a similar vein, it is essential to make an effort to develop a sense of mindfulness and presence in our professional lives. We can cultivate a greater sense of fulfillment and satisfaction in our lives by concentrating on the here and now, as well as by discovering pleasure and significance in the activities in which we are currently engaged.

In conclusion, it is abundantly clear that meaningful work brings joy. We can improve our mental and physical well-being, as well as create a life that is more fulfilling and joyful, if we are able to locate employment that is congruent with our core beliefs and sense of purpose. We can cultivate a greater sense of fulfillment and satisfaction, as well as discover joy and purpose in the work that we do, by making the practice of reflection and mindfulness a priority in our professional lives.

Chapter 20: The Happiness of Travel: The Science of Exploring the World

Exploring the world is one of the most exciting and enlightening experiences that we can have, and studies have shown that traveling can have numerous positive effects on both our mental and physical well-being. In this chapter, we will discuss the happiness that can be found while traveling as well as the science behind discovering new places.

There are many different kinds of travel, including going to unfamiliar cities and countries, discovering the natural world, and becoming fully immersed in the traditions of other countries and cultures. Studies have shown time and again that people who travel more frequently report higher levels of happiness and overall life satisfaction compared to those who do not travel.

One of the many reasons why it's so important for us to travel is so that we can gain new experiences and perspectives is because traveling gives us that opportunity. When we travel, we have the opportunity to completely submerge ourselves in new cultures, languages, and ways of life; this has the potential to inspire feelings of exploration and curiosity in us.

Similarly, going on trips can instill a sense of excitement and adventure in the traveler. We are able to experience a sense of freedom and adventure that can be invigorating and energizing when we are able to break out of our daily routines and explore new environments and experiences. When we are able to do this, we are able to feel energized.

One more reason why travel is essential to our well-being is the fact that it has the potential to instill feelings of relaxation and rejuvenation in us. We are able to feel more relaxed and rejuvenated when we are able to disengage from the pressures and requirements of our day-to-day lives and instead participate in novel and stimulating activities.

It is imperative that we place an emphasis on inquiry and investigation if we wish to broaden the scope of our lives to include more travel. This could mean scheduling regular trips or vacations, looking for opportunities for adventure and exploration in our immediate area, or simply making the effort to fully immerse ourselves in the traditions and activities of a variety of places and people.

When you go on trips, it is equally as important to train yourself to be present and mindful of your surroundings. It is possible for us to cultivate a greater sense of fulfillment and joy by concentrating on the here and now and by completely submerging ourselves in the activities and settings that are occurring all around us.

In conclusion, there is no doubt that traveling can bring happiness. We can improve our mental and physical well-being, as well as create a life that is more fulfilling and joyful, by traveling the world and immersing ourselves in the sights, sounds, tastes, and smells of other places and cultures. We can experience the joy and happiness of travel if we place an emphasis on exploration and curiosity, if we work to cultivate a sense of mindfulness and presence, and if we actively seek out opportunities for both adventure and relaxation.

Chapter 21: The Role of Genetics in Our Happiness: Are Some People Naturally Happier Than Others?

Genes are just one of many factors that can play a role in determining one's level of happiness. Happiness is an intricate and multifaceted experience that can be affected by a wide variety of factors. In this chapter, we will investigate whether or not our genes play a role in our level of happiness, as well as the question of whether or not some people are inherently happier than others.

According to the findings of some studies, our genes may be responsible for as much as half of the difference in how happy we are. This suggests that some people have a natural disposition that makes them more likely to experience higher levels of happiness than others.

The production and regulation of neurotransmitters like serotonin and dopamine, which are responsible for regulating mood and emotional well-being, is one of the primary ways that genetics can influence our level of happiness. The production and regulation of these neurotransmitters can be affected by certain genetic variations, which can have a direct impact on our happiness and well-being.

In a similar vein, our personality characteristics and coping mechanisms can be influenced by genetics, which in turn can have an effect on our levels of happiness. People who are naturally optimistic or resilient, for instance, may be better equipped to deal with stress and adversity, and they may also be more likely to experience higher levels of happiness and overall life satisfaction.

It is important to keep in mind that environmental and social factors, in addition to genetics, have a significant impact on our well-being. While it is true that our happiness can be influenced by our genes, this is not the only factor that does so. For instance, our upbringing, the social support networks that we are a part of, as well as our daily habits and routines, can all contribute to the level of happiness and life satisfaction that we experience.

Additionally, despite the fact that some people may have a natural disposition that makes them more likely to experience higher levels of happiness, it is still possible for everyone to cultivate greater levels of happiness and well-being in their lives. We can all experience greater levels of happiness and life satisfaction by prioritizing habits and practices that promote positive emotions and well-being, such as gratitude, mindfulness, and positive social connections. If we do this, we will find that our lives are more satisfying.

In conclusion, the part that genes play in determining our level of happiness is a complicated and multi-faceted one. Even though our genes might have something to do with how happy we are, our surroundings and the people in our lives also have a significant influence on our level of contentment. We can all increase the amount of happiness and joy we experience in our lives, despite the genetic predispositions we were born with, if we make it a priority to cultivate habits and practices that promote positive emotions and overall well-being.

Chapter 22: The Science of Self-Compassion: The Art of Being Kind to Yourself

The act of treating oneself with the same degree of empathy and understanding that one would extend to a trusted companion is the practice known as self-compassion. The practice of cultivating self-compassion has been shown in research to have a variety of positive effects, both on our mental and physical well-being. In this chapter, we will delve into both the academic study of self-compassion as well as the practice of treating oneself with kindness.

Self-kindness, common humanity, and mindfulness are the three primary building blocks that comprise self-compassion. Instead of being harsh with ourselves or passing judgment, practicing self-kindness involves treating ourselves with warmth and understanding. Recognizing that adversity and challenge are inevitable components of the human experience, rather than perceiving oneself as defective or insufficient, is an essential component of common humanity. Being mindful means paying attention to the here and now while maintaining an open and accepting attitude toward one's own thoughts and feelings.

The practice of cultivating self-compassion has been shown to have a variety of positive effects, both on our mental and physical health. Individuals who practice self-compassion tend to have lower levels of stress, anxiety, and depression, and they may also have better immune function and cardiovascular health outcomes, according to research.

Reframing our self-talk and the dialogue that we have with ourselves can be one way to cultivate self-compassion. We can choose to be kind and understanding toward ourselves instead of harshly criticizing or judging ourselves, and we can acknowledge that we are all flawed individuals who are deserving of love and compassion.

In a similar vein, it is essential that we make self-care and relaxation a top priority in our day-to-day lives. We can foster feelings of calm and rejuvenation, as well as cultivate a greater sense of self-compassion and overall well-being, if we make time in our schedules for rest, relaxation, and activities that promote self-care.

The cultivation of self-compassion can also be accomplished through the practice of mindfulness and awareness of the present moment. We can cultivate greater self-awareness and understanding, as well as feelings of self-compassion and acceptance, if we practice being present and aware of our thoughts and emotions in a way that does not involve judgment of ourselves or others.

In summing up, it can be said that the science behind self-compassion is crystal clear. We have the ability to increase our levels of self-compassion and overall well-being in our lives through the cultivation of self-kindness, common humanity, and mindfulness practices. We can treat ourselves with the same kindness and understanding that we would offer to a good friend if we reframe our self-talk, prioritize self-care and relaxation, and practice mindfulness and awareness of the present moment. This will allow us to create a life that is more fulfilling and joyful.

Chapter 23: The Benefits of Exercise: How Physical Activity Can Boost Happiness

Not only are exercise and other forms of physical activity important for our physical well-being, but they also have a myriad of positive effects on our mental and emotional states, which is why we should prioritize them. In the following section, we will discuss the benefits of exercise as well as the ways in which physical activity can increase one's level of happiness.

Exercising our bodies regularly has been shown to have a myriad of positive effects on our mental and emotional health. According to a number of studies, engaging in regular physical activity can have a positive impact on our mood, lessen the manifestations of both anxiety and depression, and contribute to an overall improvement in our sense of well-being.

The release of endorphins, which are naturally occurring chemicals in our bodies that promote feelings of pleasure and happiness, is one mechanism by which physical activity contributes to increased levels of happiness. Endorphins are neurotransmitters that are produced in our bodies in response to exercise. Endorphins are associated with feelings of euphoria and general well-being.

In a similar vein, physical activity has been shown to enhance feelings of accomplishment and mastery. It is possible to feel a sense of accomplishment and satisfaction when we set fitness goals for ourselves and work toward achieving those goals. This, in turn, can promote feelings of confidence and self-efficacy.

One more way in which physical activity can boost happiness is that it can serve as an outlet for negative emotions such as stress and anxiety. When we participate in physical activity, we are able to release the tension and stress that has built up inside of us, which in turn can help us feel more relaxed and at ease.

It is imperative that we prioritize physical activities that we take pleasure in and that satisfy us in order to increase the amount of exercise that we get in our daily lives. This might involve trying out new activities like dance or yoga, or finding a workout partner or group to provide motivation and support. Another option is to find a workout buddy online.

In a similar vein, it is essential to develop a sense of mindfulness and presence while engaging in physical activity. We can cultivate a greater sense of well-being and happiness in ourselves by concentrating on the here and now and devoting our full attention to the activity at hand.

In conclusion, it is undeniable that physical activity has a number of positive effects. We can improve our mental and emotional well-being, as well as our levels of happiness and overall satisfaction in life, by participating in activities that require us to move our bodies. We can make our lives more satisfying and joyful by making an activity that we take pleasure in and that satisfies us a top priority, as well as by cultivating a sense of mindfulness and presence while we are engaging in physical activity.

Chapter 24: The Happiness of Creativity: The Connection Between Art and Joy

Not only are being creative and expressing one's artistic side enjoyable ways to pass the time, but they also come with a plethora of benefits for our mental and emotional well-being. In this chapter, we will discuss the joys that come with being creative as well as the connection that exists between art and happiness.

Putting our minds and hearts to use by participating in creative pursuits can provide us with a myriad of advantages for our overall health. Research has shown that engaging in creative expression can alleviate some of the symptoms of anxiety and depression, as well as promote feelings of relaxation and well-being, and increase our overall sense of creativity and innovation.

The ability to feel a sense of self-expression and identity is one of the reasons why creative endeavors are so vital to our overall health and happiness. We are able to express our thoughts and feelings in a way that is both unique and personal when we participate in creative activities. This, in turn, can foster feelings of authenticity and aid in the process of self-discovery.

Similarly, participating in creative pursuits can increase one's sense of flow and engagement in the activity. When we are completely submerged in the process of creation, we have the potential to experience a sense of timelessness and focus, both of which have the potential to be invigorating and energizing for us.

Creativity is essential to our health for a number of reasons, one of which is that it has the potential to foster a sense of connection and community. We are able to connect with people who share our interests and passions when we participate in creative activities with other people, and this can foster a sense of belonging and connection in us.

It is essential to place a high priority on the pursuit of activities that we take pleasure in and that satisfy us in order to inject more creativity into our daily lives. This could mean learning a new artistic medium, such as painting or writing, or it could mean simply participating in creative pursuits that we find enjoyable, such as cooking or gardening.

In a similar vein, it is essential to develop a sense of mindfulness and presence whenever one is engaging in creative endeavors. We can foster a greater sense of well-being and joy in ourselves by concentrating on the here and now and completely submerging ourselves in the process of creating something new.

In conclusion, it is evident that being creative can bring happiness. We can improve our mental and emotional well-being, as well as experience greater levels of happiness and life satisfaction, if we participate in creative pursuits. We can make our lives more satisfying and joyful by giving higher priority to pursuits that we take pleasure in and that fulfill us, and by working on developing a sense of mindfulness and presence while engaging in creative pursuits.

Chapter 25: The Power of Mindset: How Our Beliefs Shape Our Happiness

To a large extent, our state of mind and the beliefs that we hold are responsible for the happiness and overall sense of well-being that we experience. In this chapter, we will discuss the power of mindset and the ways in which our beliefs can influence the level of happiness that we experience.

The way we think and the beliefs we hold can have a variety of effects on the level of happiness we experience. For instance, people who hold a growth mindset, which entails a belief in the potential for personal growth and development, have a tendency to experience higher levels of happiness and life satisfaction than those who hold a fixed mindset, which entails a belief that our qualities and abilities are fixed and unchangeable. This is because individuals who hold a growth mindset tend to believe that they can grow and develop personally.

In a similar vein, the beliefs we hold about our own capabilities and potential can have an effect on the degree to which we experience happiness and well-being. People who believe they are unable to make meaningful changes in their lives are more likely to be unhappy and dissatisfied with their lives overall. For instance, people who believe they are capable of achieving their goals and overcoming obstacles are more likely to experience high levels of happiness and satisfaction in their lives.

The way we think and the beliefs we hold can also have an effect on our capacity to deal with pressure and difficulty. Individuals who hold a resilient mindset, which involves a belief in our ability to overcome adversity and bounce back from setbacks, tend to experience lower levels of stress and anxiety, and may be more likely to experience greater levels of happiness and life satisfaction. This is because a resilient mindset involves a belief in our ability to overcome adversity and bounce back from setbacks.

It is important to prioritize practices and habits that promote positive thinking and self-belief in order to cultivate a positive and growth-oriented mindset. This can be accomplished by reading and listening to positive and uplifting content. This could involve engaging in activities such as practicing gratitude and positive self-talk, looking for opportunities for personal growth and development, and recasting unhelpful thoughts and beliefs as more constructive and empowering narratives.

In a similar vein, it is essential to make a concerted effort to develop a sense of mindfulness and presence in our day-to-day lives. We are able to identify and challenge beliefs that are negative or limiting, and promote a mindset that is more positive and growth-oriented when we are aware of our own thoughts and beliefs in a way that is non-judgmental about them.

In conclusion, the power of one's mental attitude cannot be denied. We have the ability to improve our levels of happiness and well-being, as well as triumph over the challenges and obstacles that arise in our lives, if we cultivate a growth-oriented and growth-focused mindset. It is possible to live a life that is more satisfying and filled with joy if we make it a priority to engage in behaviors and routines that encourage optimistic thinking and self-belief, and if we work to develop a sense of mindfulness and presence in our day-to-day activities.

Chapter 26: The Joy of Learning: The Connection Between Knowledge and Happiness

Learning new things and expanding our existing knowledge not only contributes to the development of our intellect, but it also has a wide range of positive effects on our mental and emotional well-being. In this chapter, we will discuss the pleasures associated with learning as well as the link that exists between education and contentment.

Learning new things has been shown to have a variety of positive effects on both our mental and emotional well-being. Learning throughout one's life has been shown to have a positive impact on cognitive function, a reduction in the symptoms of depression and anxiety, as well as an enhancement of our overall sense of well-being and satisfaction with life.

The development of a sense of curiosity and wonder is one of the ways in which education can help to increase one's level of happiness. It is possible to cultivate a sense of intellectual curiosity and exploration by participating in new learning opportunities, which give us the opportunity to investigate new fields of study and concepts.

Similarly, education can foster feelings of accomplishment and mastery in its students. It is possible to feel a sense of accomplishment and satisfaction when we set learning goals for ourselves and work toward achieving those goals. This can contribute to increased feelings of self-confidence and self-efficacy.

One more way that education can increase one's level of happiness is by opening doors to new possibilities for social connection and community. When we participate in learning opportunities with other people, we increase our chances of making connections with other individuals who share our interests and passions, which in turn can foster a sense of belonging and connection.

In order to incorporate more learning into our lives, it is important to prioritize opportunities that we find both enjoyable and fulfilling. This will allow us to increase the amount of learning that we incorporate into our lives. This could involve enrolling in a course or workshop on a new subject or topic, or it could simply involve participating in opportunities for self-directed learning, such as reading or exploring new subjects online.

In a similar vein, it is essential to develop a sense of mindfulness and presence whenever there is an opportunity to learn something new. We can foster a greater sense of well-being and joy in ourselves by concentrating on the here and now and by completely submerging ourselves in the process of learning.

In conclusion, there is no denying the sheer delight that comes from learning. By taking advantage of opportunities for lifelong learning, we can improve our mental and emotional well-being, as well as experience higher levels of happiness and a greater sense of satisfaction with life. We can make our lives more meaningful and enjoyable by giving higher priority to activities that bring us pleasure and satisfaction and by cultivating a sense of mindfulness and presence while engaging in educational pursuits. This will allow us to create an existence that is more satisfying and joyful.

Chapter 27: The Happiness of Music: How Sound Can Boost Our Well-Being

Not only can music provide us with a source of entertainment and pleasure, but it also has a great many positive effects on our mental and emotional health. In this chapter, we will discuss the factors that contribute to happiness, specifically music and sound, as well as the connection between the two.

The mental and emotional health of those who listen to music regularly can be improved in a variety of ways. Music has been shown to alleviate symptoms of anxiety and depression, promote feelings of relaxation and well-being, and increase our overall sense of creativity and productivity, according to research that has been conducted.

The release of dopamine, which is a naturally occurring chemical in our brains that promotes feelings of pleasure and happiness, is one mechanism by which music can be said to increase levels of happiness. When we listen to music that we enjoy, dopamine is produced in our brains. Dopamine is a neurotransmitter that is associated with feelings of euphoria and general well-being.

In a similar vein, listening to music can help foster a feeling of connection and community. When we listen to music with other people, we have the opportunity to connect with people who share our musical interests and preferences. This can help to foster a sense of belonging and connection in the listeners.

One more way that listening to music can improve one's mood is that it can evoke feelings of nostalgia and bring back fond memories. It is possible for us to feel feelings of comfort and happiness when we listen to music that is associated with positive memories or experiences. This can be especially beneficial during times of stress or difficulty, as it can help us feel less stressed and more happy.

It is essential to give precedence to music that we take pleasure in listening to and that satisfies us in order to make room for more music in our lives. This might involve playing music in the background while we go about our daily activities, going to live performances of music such as concerts or other events, or even learning to play an instrument or sing.

During musical experiences, it is equally essential to develop a sense of mindfulness and presence in order to get the most out of them. We can promote a greater sense of well-being and joy in ourselves by completely submerging ourselves in the music and paying attention to the sounds and feelings that it conjures up in us at the same time.

In conclusion, there is no doubt that music can make people happy. We can improve our mental and emotional well-being, as well as experience greater levels of happiness and life satisfaction, simply by listening to music that we find enjoyable and that satisfies us on a personal level. We can make our lives more meaningful and fulfilling by making music a higher priority as a source of entertainment and enjoyment and by cultivating a sense of mindfulness and presence during musical experiences. This will allow us to feel more in the moment.

Chapter 28: The Science of Mind-Body Connection: How Our Physical Health Affects Our Happiness

The state of our mental and emotional health as well as our physical health and well-being are inextricably linked to one another. In this chapter, we will investigate the scientific research behind the mind-body connection and the ways in which our physical health can influence the level of happiness that we experience.

According to the findings of various pieces of research, the state of our bodies can have a significant influence on our mental and emotional well-being. For instance, people who engage in regular physical activity and maintain a healthy diet have a tendency to experience lower levels of stress and anxiety, and they may be more likely to experience greater levels of happiness and life satisfaction. These benefits can be attributed to the positive effects of lifestyle choices.

In a similar vein, the state of our physical health can have an effect on how well we are able to deal with pressure and difficulty. People who live healthy lifestyles may be better able to deal with stressful situations and anxiety, as well as be more resilient in the face of difficult circumstances.

The release of endorphins, which are naturally occurring chemicals in our bodies that promote feelings of pleasure and happiness, is yet another way that our physical health can have an effect on our level of happiness. Endorphins are neurotransmitters that are produced in our bodies in response to exercise. Endorphins are associated with feelings of euphoria and general well-being.

It is essential to put an emphasis on behaviors and routines that improve one's physical health and well-being if one wishes to foster a stronger mind-body connection, as well as to increase one's level of happiness and overall sense of well-being. This could entail going to the doctor when necessary, maintaining a healthy diet and sleep schedule, and engaging in regular physical activity as part of the routine.

In a similar vein, it is essential to incorporate practices of mindfulness and presence into our day-to-day lives and to be conscious of the ways in which our physical health can have an effect on our mental and emotional well-being. We have the ability to craft a life that is richer in meaning and filled with joy if we take a holistic approach to our health and well-being.

In conclusion, there is no doubt that there is a connection between the mind and the body. We can boost our levels of happiness and well-being, as well as experience greater levels of resilience and the ability to cope with adversity, if we make it a priority to engage in behaviors and routines that are beneficial to our physical health and well-being. We can make our lives more meaningful and enjoyable by developing a sense of mindfulness and presence in our daily activities, and by taking a holistic approach to improving our health and well-being.

Chapter 29: The Happiness of Authenticity: The Importance of Being True to Ourselves

For the sake of our mental and emotional health, it is essential for us to maintain our authenticity and honesty with ourselves. In this chapter, we will discuss the happiness that can be found in authenticity as well as the significance of remaining true to who we are as individuals.

When we are able to express our true thoughts, feelings, and values in a manner that is honest and genuine, we can say that we are being authentic. This can foster feelings of self-discovery and identity, as well as a sense of self-worth and confidence, and it can promote feelings of empowerment.

In a similar vein, being honest with ourselves can enhance our feelings of connection and belonging to a community. We are better able to foster a sense of community and support for one another when we are able to express our genuine selves and connect with others who share our beliefs, values, and interests.

It is essential to put practices and routines that encourage self-awareness and self-discovery at the forefront of one's life if one aspires to be more genuine and honest with oneself. This could involve writing in a journal, meditating, or seeking professional help in the form of counseling or therapy.

In a similar vein, it is essential to make a concerted effort to develop a sense of mindfulness and presence in our day-to-day lives. We are able to recognize and question beliefs that are unhelpful or limiting, which in turn paves the way for an existence that is more genuine and satisfying when we do so without passing judgment on our own thoughts and feelings.

In conclusion, it is crystal clear that authenticity leads to happiness. We can improve our mental and emotional health, as well as experience higher levels of self-worth and confidence, if we are true to ourselves and express our genuine thoughts, feelings, and values in a way that is authentic and honest. We can make our lives more meaningful and enjoyable by making the formation of practices and routines that encourage self-awareness and self-discovery a top priority, as well as by working to develop a sense of mindfulness and presence in our day-to-day activities.

Chapter 30: The Joy of Giving Back: The Connection Between Community Service and Happiness

Not only is it crucial that we give back to our communities and assist those in need for the sake of the greater good, but doing so also has a wide range of positive effects on our mental and emotional well-being. In this chapter, we will discuss the satisfaction that comes from helping others and the link that exists between volunteering and overall happiness.

Volunteering and community service give us the opportunity to have a constructive influence on the lives of others; as a result, we may experience increased feelings of self-worth and a sense of direction in our lives. Helping other people can also cultivate a feeling of connection and community because it requires us to collaborate with other people in order to achieve a common objective.

Research has shown that people who regularly participate in volunteer work and community service are more likely to report higher levels of happiness and overall life satisfaction than those people who do not engage in such activities. Oxytocin is a naturally occurring chemical in our bodies that fosters feelings of social bonding and overall well-being. It is possible that the release of oxytocin is responsible for at least some of this effect.

In order to make more time for volunteer work and community service in our lives, it is important to find causes and organizations that are congruent with the things that are important to us and the things that we are passionate about. This could involve helping out at a community center near where you live, giving money to a charity or other non-profit organization, or even just being kind and generous to other people.

In a similar vein, it is essential to develop a sense of mindfulness and presence in order to make the most of one's time spent volunteering or serving the community. We can recognize and question beliefs that are unhelpful or limiting, which paves the way for a life that is more satisfying and brings us more happiness if we practice mindful awareness of our thoughts and feelings without passing judgment on them.

The satisfaction that comes from being able to help others is undeniable. Through participation in volunteer work and community service, we have the opportunity to make a positive difference in the lives of others, foster feelings of self-worth and purpose, and experience higher levels of happiness and overall life satisfaction. We can cultivate a life that is more satisfying and full of joy by giving our attention and resources to causes and organizations that are congruent with our beliefs and pursuits, as well as by practicing mindfulness and being fully present while performing acts of community service and volunteer work.

About the Publisher

Accepting manuscripts in the most categories. We love to help people get their words available to the world.

Revival Waves of Glory focus is to provide more options to be published. We do traditional paperbacks, hardcovers, audio books and ebooks all over the world. A traditional royalty-based publisher that offers self-publishing options, Revival Waves provides a very author friendly and transparent publishing process, with President Bill Vincent involved in the full process of your book. Send us your manuscript and we will contact you as soon as possible.

Contact: Bill Vincent at rwgpublishing@yahoo.com

www.ingramcontent.com/pod-product-compliance
Lightning Source LLC
LaVergne TN
LVHW041649060526
838200LV00040B/1769